# FIRE!
## AT WAR WITH THE RED DEVIL

# FIRE!
## AT WAR WITH THE RED DEVIL

TEXT AND
PHOTOGRAPHS BY
**THOMAS K. WANSTALL**

ADDITIONAL TEXT BY
**GEORGE HALL**

PRODUCED BY
**BECKY S. CAVANAUGH**

**BONANZA BOOKS**
**NEW YORK**

This 1989 edition is published by Bonanza Books,
distributed by Crown Publishers, Inc., 225 Park Avenue South, New York, New York 10003.

Printed and bound in Hong Kong


Library of Congress Cataloging-in-Publication Data

Wanstall, Thomas K.
Fire! at war with the red devil / text and photographs by Thomas K. Wanstall.
1. Fire extinction—Pictorial works.   2. Fire departments—Pictorial works.
3. Fire fighters—Pictorial works.   I. Wanstall, Thomas K.   II. Title.
TH9310.5.H34     1989
628.9′2—dc19                                                                 89-714
                                                                              CIP

ISBN 0-517-67953-1
h g f e d c b a

# FORMULA FOR FIGHTING FIRE

**B**enjamin Franklin, a champion of volunteer fire fighting in Colonial America, said it best when he called fire "a good servant but a terrible master." Man has struggled to control fire ever since prehistoric times, and the rules for doing so are very much the same as they were in a Cro-Magnon cave.

Fire is easy to recognize but difficult to define. It is a plasmalike phenomenon that results when the temperature around a combustible material is elevated to a certain level. In a typical wood-frame dwelling, that temperature is around 400° F. The wood will vaporize, throwing off grayish brown fumes (smoke); visible flames will appear at the points of vaporization.

We remember from grade-school science class that fire needs three things to perpetuate itself: material that will burn, oxygen, and enough heat to initiate the reaction. This is known as the "fire triangle." Remove any of these elements and the fire will quickly die. The best mechanism for achieving this end is the same as it has always been—douse the flames with plenty of water. Water is an amazingly efficient ingredient in fire suppression. It soaks up heat better than any element on earth; even when it turns into steam, at approximately 212° F, it remains far below the ignition point of most combustible materials. And water also separates fire from the air it needs to survive.

Water, however, is not always the answer to fighting fires. Often fires of a chemical or electrical nature must be attacked from a different side of the fire triangle. For example, use of water on liquids such as gasoline or oil would cause the fire to flow, thus creating additional danger of spreading the flames. Also, many burning chemicals become lethal when exposed to water. When fire fighters are faced with these problems, they turn to foam and fire-retardant powders, thereby attacking a different side of the fire triangle. Starving the fire of oxygen is as effective as cooling it with water.

Modern fire-engine pumpers are capable of throwing as much as 1,500 gallons of water per minute through several hose lines onto relentless flames. The emphasis, of course, is on getting water onto the flames as rapidly as possible. The engine's driver, also known as the chauffeur or the engineer in some departments, is the person responsible for connecting the rig to a water source immediately upon arrival at a fire. In most cities that source is a street-corner hydrant connected to a high-pressure water system specially geared for fire fighting. The engine can also drop a suction line into an underground cistern or draft directly from the virtually unlimited water supply in a river, harbor, or lake. It is not always that simple, however. In many rural areas, large water supplies are not readily available. In these areas, volunteer fire fighters actually bring their water with them. This is accomplished by the use of huge tanker trucks designed especially for fighting fires. A complicated routine of shuttling several tankers to a water source and back to the fire is carried out like a precision military drill. Not to be forgotten, too, are residential swimming pools. Many a fire has been nipped in the bud, or at least held at bay until additional help arrives, by drafting directly from a swimming pool.

While the engine men are at work dousing the fire with water, other fire fighters will be attacking it from a different direction—lowering the temperature around the fire by ventilating the structure. By breaking out windows and cutting holes in the floors and roof above the flames, the fire fighters help it to breathe and cool itself down faster. This procedure allows the superheated gases emitted from burning contents to escape before they literally explode. Bystanders often misinterpret this frenzy of glass breaking and chain-sawing as unnecessary destruction, but the fire fighters know exactly what they are doing. Their efforts will most likely save the structure from ruin.

# PUSHIN' IN—The Engine Company

Fire fighting is a team effort. The expertise of fire-fighting personnel and the special capabilities of their equipment are most successfully employed in a carefully coordinated and rehearsed operation. At a typical working fire, the fire fighters will arrive aboard several specialized pieces of apparatus. Let's take a closer look at the principal types of rigs in common use today.

The fire engine is the vehicle that allows fire fighters to pump water onto the fire. When the engine has arrived at the fire scene and has been connected to a hydrant by a "jumper," its engine is disengaged from the drive shaft and linked to a powerful internal pump that pushes water from the hydrant through various hose lines. Depending on the

volume of fire, the engine's officer will have selected an appropriate hose size and nozzle. He and his hose team will then move in on the seat of the fire, crouching behind the high-pressure spray from the "knob." It is a long-standing tradition of engine men that being "on the knob" is the only place to be when "pushin' in." The fury of a hot, rolling fire mandates relief for the knob man every few minutes. The engine officer pushes right in with the fire fighters. His experience and knowledge play a crucial role in battling the fire. The catastrophic possibility of a "flash-over" or "flashback" is always present, and it is the officer's responsibility to be alert and protect his company from this danger.

A flashover occurs when a burning room

and all of its contents become superheated to the point that everything in the room, from the floor to the ceiling and from wall to wall, explodes into a raging inferno. This can happen when a new source of oxygen is introduced into a fire. Animated like a hungry tiger, the fire will make a rush toward this new source of energy. The fire will swoop down long hallways in an instant, destroying anything in its path. Veteran fire fighters will tell you, "You know when it's comin' when you feel a cool breeze on the tips of your ears. Then it's time to hit the floor."

Most modern fire engines are equipped with an internal booster tank, which is usually kept loaded with about 500 gallons of water. This water source can be used with a small-diameter hose reel for attacking smaller fires, such as brush and rubbish fires, that don't warrant a hydrant hookup. In a desperate emergency, the hose team can use this water for a lifesaving initial attack. This is a risky move, however; the booster will only last a couple of minutes, and the engine's pump operator must hook up to a hydrant quickly or risk stranding his fellow engine men inside a burning hall or room with a dry line.

For the big burners—the fully involved buildings, which are unsuitable for an interior attack—many engines are equipped with the big guns. The "deck gun" is a large, permanently affixed, directional nozzle mounted atop the apparatus. It has the capability of delivering large volumes of water into a burning building from the exterior. Along with the deck gun, many engines carry a portable "deluge set." Similar to a deck gun, the deluge set can be removed from the apparatus and strategically located in rear yards or alleys to provide large-caliber streams (known as master streams) to areas otherwise inaccessible to the fire engine.

# TRUCKEES—True Grit

While the engine crews worry about putting water on the flames, the fire fighters of the truck and ladder companies—affectionately known as truckees—attend to a completely different set of priorities. Fire trucks come in various sizes and shapes. Some have hydraulic aerial ladders that will extend a hundred feet or more, while some have elevating platforms that will rise to a similar height. Both can be quickly set up for window rescues or to remove occupants from fire escapes. They are also used for placing fire fighters on roofs with axes and saws. The principal responsibilities of a truck company are to search for and rescue building occupants and to ventilate the burning structure to hasten the exodus of heat and smoke. To accomplish all this, the truck carries an assortment of hand-erected ladders of differing heights, as well as some very specialized ladders for use on peaked roofs and for scaling otherwise impossible-to-reach areas.

Truckmen also have the unenviable responsibility of overhauling the stricken building after the fire—that is, searching for lingering hot spots and salvaging property still undamaged by heat, smoke, and water. Trucks carry the necessary tools for this tiring, painstaking work. Pole-mounted steel hooks are used for pulling apart ceilings, while forcible-entry tools are used to pry open moldings around windows and doors in search of hidden fire. Power fans and pumps are used for expelling smoke and water.

Truckmen also throw their share of water at large blazes. Aerial ladders fitted with high-volume ladder pipes can be used to hit the fire from above. The elevating platform—known as an aerialscope, tower ladder, or ladder tower—has become a very popular truck in many departments. Many of these towers will rise more than 100 feet into the air, allowing for an extremely effective aerial attack. The "bucket" can be erected very quickly at the fire scene, and the truckmen can move from window to window to execute rescues or throw an accurate stream of water from the permanently mounted deck gun.

# SMALL TOWNS AND VOLUNTEERS

Volunteer fire fighting has a long and distinguished tradition in American history. The first organized fire-fighting efforts in the United States were strictly volunteer affairs, and even the biggest American cities like Philadelphia and New York relied completely on volunteer companies until later in the nineteenth century, when the first paid, professional fire fighters made their appearance. The volunteer firehouse was the seat of considerable political and cultural power in American cities before the Civil War; the notorious Boss Tweed, corrupt overlord of New York City politics in the 1870s, built and ran his political machine from a volunteer firehouse on the Lower East Side, where he served as captain of a hose company.

Today, thousands of small communities rely completely on volunteer companies for fire protection, as well as for emergency medical assistance. Most of these "vollies" are every bit as well trained and well equipped as their paid, big-city counterparts; indeed, some of the volunteer outfits field the most spectacular and up-to-date apparatus to be found anywhere. Less affluent departments make do with refurbished equipment retired from active duty in metropolitan areas.

Often the volunteer firehouse is the social center of a community. Fund-raising activities are conducted to provide for new equipment, and weekly drills are held so that volunteers stay sharp and their fire-fighting equipment remains well-maintained. In recent years, demands on volunteers have increased dramatically. Sleepy little villages awaken to find an interstate highway running through their backyards, changing life in the firehouse altogether. Suddenly they are responding to multivehicle accidents with oil and chemical spills. Populations quadruple as these once-quiet towns become bedroom communities for metropolitan workers, and, in turn, the number of fire calls increases. The fire chief, whose worries were relatively few, must now keep current with the latest fire-fighting techniques and consider adding to his fleet the newest, most sophisticated equipment. Budgets that were once provided by car washes and bake sales now require huge allocations from municipal coffers. The volunteer fire fighter—for whom the big burner was a rare occurrence—suddenly faces a potentially dangerous situation every time the alarm sounds.

In volunteer departments, the fire fighters are summoned by radio paging equipment in their homes and portable pagers carried on belts or in pockets, and by the traditional alarm at the firehouse. Some members respond to the firehouse in order to man the fire engines, while others respond directly to the scene. Well-trained and motivated volunteer companies in this configuration can turn in response times very close to those of professional departments. Because of rapid growth, many communities require full-time fire fighting coverage but find it impractical and costly to maintain a full-time, paid fire department. Their solution to this dilemma is to provide paid fire fighters to drive the apparatus to the scene immediately upon receiving an alarm. Some departments also provide a small crew—two or three additional fire fighters—to operate a single piece of equipment. This crew is supplemented by volunteers, who respond with additional apparatus and manpower.

We often hear about the hair-raising exploits of the big-city fire fighters responding to devastating and dramatic fires. The efforts of small-town fire departments, just as heroic, are often overlooked by the public. Consider for a moment the town of Ardsley, New York, where there hadn't been a major fire in more than twenty years. Routine calls to the New York State Thruway had been keeping the volunteer department busy when an alarm was received from a recently constructed warehouse in the town center. Upon arrival, the firefighters were confronted with a fully involved building requiring all the resources of their town and surrounding communities. It was a blaze matching any you might find in a big city.

Further, consider a suburban New Jersey fire department's problems while returning from a routine investigation on a nearby interstate. In this department, consisting of professionals and volunteers, the on-duty paid crew sounds the alarm for the volunteers, notes the location of the alarm, and responds with the engine to the troubled site. Subsequent manpower and apparatus are quickly dispatched by the volunteers. Radio communications are handled by a preassigned volunteer who arrives at the firehouse minutes after the initial alarm. This system had worked well for years and under ordinary circumstances will continue to work well, but fire fighting is not an ordinary business. On this occasion, while returning from the investigation in the early hours of the morning, the crew followed a smoke odor into an isolated industrial area. Here they found a fully charged structure in need of immediate attention. In a big city, a quick radio transmission would rapidly bring three engines, two trucks, and approximately twenty-five men. But here was a crew of only three who had to do everything. One man vented the windows while another looked for trapped victims. The third man held back the flames by judiciously using the water in the booster tank (normally the second engine to arrive wraps a hydrant and relieves the first engine crew). At the same time he eyeballed a hydrant, figuring he might have to use it in the event help did not arrive. The physical demands on a fire fighter during the initial stages of a fire are enormous, and what was, in reality, only minutes passing seemed like hours. Just as their strength and water supply began to ebb, three shining rigs turned the corner with a full complement of volunteers. Though pushed to the limit, the system worked—a phone call from a citizen had alerted the volunteers. The professional fire fighters were exhausted, and the arrival of the volunteers was a welcomed sight. Clearly, the danger experienced by courageous fire fighters is not limited to the infernos of the city. Often the hero turns out to be the person next door.

# DISPATCH

A fire can develop and grow with frightening speed, and the job of the fire fighter is always to respond as rapidly as possible to effect rescues and to attack the seat of the fire. The men in the firehouse rely on the dispatchers, usually located in a windowless communications center elsewhere in the city or county, to field calls from citizens and alarms from street-corner boxes and then put the appropriate companies onto the streets as quickly as possible.

Pull boxes on corner light poles were once the principal means of summoning fire fighters, and some cities still maintain these antiquated firebox systems, although they tend to be a very troublesome source of false alarms. Many cities now use two-way voice systems that allow citizens and dispatchers to speak with one another.

Whether by box or phone, the incoming call will be interpreted by the dispatcher and the necessary action taken. Most modern departments in big cities have a computer-aided dispatch system, with all of the city's addresses keyed into its memory, along with corresponding information on cross streets, potential hazards, or unusual building materials, as well as a listing of the engines, trucks, squads, and chiefs that are expected to respond to that location.

Some incidents will require only a single engine or a medical unit along with a fire company as backup. If the phone call reports a car fire or a rubbish fire, usually only an engine or an engine and truck will be assigned. If the phone call reports a building or structure of any kind on fire, the dispatch will be a "full box," or a first-alarm assignment; to which two or three engines, one or two trucks, and one or more chiefs are sent. It takes a fairly serious fire to get past a first-alarm response; with a big "worker," chiefs at the scene will radio for a second, third, fourth, or fifth alarm. A five-alarm fire in a city the size of San Francisco or Boston will put half of the on-duty fire fighters at the scene. With so many personnel and pieces of equipment responding to the greater alarm, the dispatchers will move the remaining companies around the city to even out the fire coverage in the event of another major fire. Most municipalities also maintain mutual-aid agreements with nearby departments; these companies can be called across city lines to pitch in when one or more fires threaten to overwhelm the city's fire-fighting forces.

Civilians are trained to be dispatchers in many cities, while other departments prefer to have experienced fire fighters manning the telephones and computer consoles. In either case, the person who answers a call for help will be well equipped to take quick and effective action on the caller's behalf.

# LEARNING TO BE A FIRE FIGHTER

**W**ould-be fire fighters have much to learn before they try to push a hose through their first smoke-filled hallway or attempt to carry a victim down an extended aerial ladder. Much of the material to be absorbed is technical—the physics of fire, the behavior of fire, and the tactics of its suppression. It is not just a matter of dousing the flames with water. Endless hours are spent in the classroom learning such things as hydraulics—pounds of pressure per square inch, gallons per minute, and friction loss as water passes through the hoses. Also taught is the proper handling of myriad specialized fire-fighting tools.

Classroom accomplishments are put to the test in the smoke chambers and mock structures of the average training center. Hours and hours are spent crawling through hot, smoky rooms searching for mock victims and climbing fully extended aerial ladders in order to "rescue" people from windows. Fabricated propane gas fires are extinguished, along with oil-pit fires, which spew black smoke, heat, and flames. Driver training—learning to maneuver the cumbersome rigs through crowded streets—and extensive first-aid training round out the approximately ten weeks of study required to be a rookie fire fighter.

The fledgling trainee will also find that he has to face and control his understandable fear. No fire fighter enters a burning structure without being fully aware of the risk involved. Fighting fires can be enormously claustrophobic, with clumsy, restrictive air masks, confusing sounds and shouts coming from all directions, and choking smoke that reduces visibility to almost nothing. A fear of heights must also be mastered if the fire fighter is going to be able to function safely and effectively on a rain-slick pitched roof or at the swaying tip of a 100-foot aerial ladder.

Even young trainees in good physical condition are astonished at the amount of strength and stamina required by the job. Fire fighting is a blur of dragging water-filled hoses, carrying heavy gear up flight after flight of stairs, and flailing away with axes, all the while wearing as much as forty pounds of protective gear. Only men and women in top condition will be able to pass the grueling physical-fitness tests given by many fire departments across the country.

# HAZARDOUS-MATERIALS UNITS

"Haz-mat," short for hazardous-materials handling, is a term unheard of in fire fighting a generation ago. Increased awareness of the dangers of toxic chemicals and the peculiar problems of their neutralization and disposal has led to this new branch of fire-fighting technology. Many sizable departments maintain a fully manned haz-mat truck equipped with chemical-testing gear, protective apparel, an assortment of chemical containers, and absorbing pads for the removal of toxic materials. Many of these units are equipped with modular telephones and computer consoles to ensure immediate communication with information sources concerning unidentified materials.

Fire fighters in a hazardous-materials unit take on a difficult and dangerous job, where one accident or mistake could be fatal or cause serious health complications.

# HEAVY RESCUE AND SPECIAL APPARATUS

No other piece of fire equipment can turn heads on a city street more quickly than a fire rescue truck. Because of its awesome size and vast array of sirens and lights, you just know it's on its way to something big. Fire fighters of the rescue company are considered the elite—a designation they deserve.

In big cities, one or more of these units are sent to every working fire. It is their duty to search for and rescue trapped victims, whether they are civilians or fire fighters, causing one veteran member of a rescue unit to boast, "The fire fighters save the people, and the rescue saves the fire fighters."

Rescue units also respond to all types of life-threatening situations: building collapses, major vehicular accidents, subway and tunnel fires, and underwater (SCUBA) searches and high-altitude rescues. The list of equipment carried on those magnificent trucks seems endless—generators; cables; portable power tools and inflatable air bags for lifting extreme weights; Hurst tools (jaws of life) for prying open mangled vehicles; acetylene torches; rappelling gear; and heat-seeking sensors that detect fires are all found on board.

# THE BIG GUNS

In the late sixties, many of the older, big cities found themselves confronted with huge conflagrations in aging factories and tenements. Conventional interior fire fighting was out of the question. Huge volumes of water needed to be applied from a reasonably safe distance. In places like New York, Chicago, and Philadelphia, engineers and mechanics combined their wisdom and came up with apparatus like New York's super pumper system and Chicago's "Big John." Big John, conceived and constructed in-house by the Chicago Fire Department, sported hydraulically operated nozzles that could toss up to 3,000 gallons of water per minute. New York's super pumper was built by the Mack Truck Company. The world's largest fire engine, it was put into service in 1965 and retired in 1982. Designed as a tractor trailer, the 2,400-horsepower pump could push out as many as 8,800 gallons of water per minute, feeding eight standard pumpers or three of its satellites.

With the retirement of the super pumper, a more flexible system, the maxi-water system, was put into place. Each of six satellites is accompanied by a 2,000-gallons-per-minute pumper and supplemented with a 1,000-gallons-per-minute, high-pressure pumper for use in high-rise buildings.

# LIFE IN THE FIREHOUSE

Fire fighting is unlike other jobs. There is an element of shared danger. Burns, smoke and toxic-gas inhalation, lacerations from broken glass, sprains and fractures from falls down ruined stairways or from collapsing debris, and a hundred other potential disasters lurk behind every call.

Another unique aspect of a fire fighter's working life is the sheer amount of time he or she spends with colleagues. In most paid departments, the fire fighters work twenty-four hour shifts with forty-eight hours off; some cities prefer a system of mixed nine- and fifteen-hour shifts. In either case, the men and women who fight fires actually live together as well as work together. The typical fire fighter spends as much time in the firehouse with his co-workers as he does with his family at home. It's no wonder that a special bond develops between the practitioners of this hazardous and noble profession.

Firehouse life is a mixture of dramatic excitement and long hours awaiting the next call. Most professional fire companies set aside time every day for performing drills and practicing medical and first-aid skills. The companies also tour their district, inspect individual structures, and familiarize themselves with every square inch of their neighborhood. Knowing the way around a structure and being familiar with any potentially dangerous substance that may be stored inside can make all the difference in fighting a fire.

Since firemen spend so much time together, their firehouse becomes a second home for them, and its maintenance is their responsibility. Housekeeping, referred to as committee work, also extends to their gear and vehicles, which can't be cleaned and checked often enough. After a real working job, there are hose lengths to be cleaned and dried, booster tanks to be refilled, and many other chores to finish before the rigs are restored to fighting trim.

And then there's the matter of firehouse cuisine. Fire fighters are famous for eating well on the job. Everyone is expected to take a crack at cooking now and then, but in most firehouses the best chefs tend to the serious cookery, while the others handle the dishwashing. Contrary to popular belief, the city doesn't pick up the dinner tab—fire fighters do their own grocery shopping and pay their own food bills. There's always one culinary fiasco the fire fighters can't control—the frustrating likelihood that their carefully prepared meal will be ruined when the alarm sounds just as they take their seats at the table. One requirement of a successful firehouse recipe, therefore, is that it taste just as good reheated. Many a firehouse dinner has been lovingly prepared but not eaten until the early hours of the morning.

# FUN AND GAMES

A fire muster is a gathering of fire fighters and citizens who support their fire department's efforts. At a muster you'll find antique fire trucks, period uniforms, helmets, and other pieces of antiquated equipment. The high point of the day comes when the antique fire engines are put through their paces, demonstrating that they still can pump just as well as they did in their prime.

Companies also compete against each other in the fine art of stretching hose and hitting targets, ascending ladders, and, of course, presenting the best-looking rig.

It's an all-day affair that allows fire fighters to acknowledge their profession's noble heritage.

# PUMPER
## Mack-Ward 79

Engine: Mack Turbo-charged 350 hp
Transmission: Allison 4-speed automatic
Overall Length: 317″
Wheel Base: 166″
Pump: Aqueous 2-stage 1,000 gpm
Hose: 28 lengths• of 3½″ hose
(1) 5 lengths of 2½″ hose
(1) 5 lengths of 1¾″ hose
(1) 3 lengths of 1¾″ hose
200′ 1″ booster hose
Equipment
 Oxygen tank for first aid
 (2) Federal Night Fighters—600,000 candle power spot/flood lights
 (2) 10′ 4½″ hard suction hoses for drafting
 24′ extension ladder

•Each length is 50 feet

# TOWER LADDER
## Mack 75 Aerialscope Tower Ladder

Engine:  Mack Turbo-charged 350 hp
Loaded Weight:  56,000 lbs.
Road Speed:  56 mph
Wheel Base:  240"
Working Height:  75'/75°
Transmission:  Allison 4-speed automatic
Platform Capacity:  1,000 lbs.
Platform Control:  Single stick with dead-man switch electrically actuates
    hydraulic control valves located at turntable for position control.
    Platform automatically maintains a level position through all boom
    movements.
Discharge:  1,000 gpm via an Akron 3426 deck monitor with stack tips 1¼", 1½", and 2"

Equipment

Portable ladders (8)—from a 10′ folding ladder to a 35′ extension ladder

Pike poles (8)—4 six-foot, 1 eight-foot, 2 ten-foot, 1 twelve-foot

Halligan Hooks—two

Electrical smoke ejector

Forcible entry tools

Air bags

4 one-hour air masks for high-rise operations

Roof ropes

Stokes

Tarps

Cutting torch

## AERIAL LADDER

Seagrave 100' Aerial Ladder
Gross Weight: 40,000 lbs.
Wheel Base: 226"
Road Speed: 55 mph
Working Height: 100'
Engine: Detroit diesel 6v92 350 hp
Transmission: Allison 4-speed automatic
Ladder Control: Electro-hydraulic controls for extension
    and rotation from the turntable, manual throttle. Maximum time
    permitted for each 360° rotation—50 seconds. Minimum time
    permitted for each 360° rotation—40 seconds. With 200-lb.
    weight must be capable of being fully extended, fully elevated,
    and rotating 90° within 1 minute.
Ladder Pipe: Akron 1496 3 tips (1¼", 1½", 1¾")
Discharge: 500 gpm
Equipment:
    Pike poles (8)—4 six-foot, 2 eight-foot, 1 ten-foot, 1 twelve-foot
    1 eight-foot hook and 1 six-pound ax permanently mounted on aerial ladder
    Axes—6 six-pound, 6 eight-pound
    Lucas extraction tool
    Federal Night Fighters (2)—600,000 candlepower spot/flood lights
Ground Ladders (8)—From 10' folding ladder to 35' extension ladder

# RESCUE TRUCK
**Mack R-model rescue van with Pierce body**

Empty weight:  26,400 lbs.
Engine:  Mack Thermodyne Diesel 6-cylinder 275 hp
Transmission:  Allison 4-speed automatic

## ENGINEMAN

1. Leather Cairns helmet
2. Nomex turnout coat
3. Handie-Talkie speaker/microphone
4. Hose
5. Thigh-high protective rubber boots
6. Apartment pack (contains 150' lightweight 1¾" hose, reducer, automatic nozzle, hydrant wrench, spanner)
7. Air mask (connects to tank worn on back; positive pressure mask rated at 30 minutes)
8. Protective gloves

ALL GEAR IS OSHA APPROVED

## TRUCKMAN

1. Leather Cairns helmet
2. Nomex turnout coat
3. Protective gloves
4. Ax and forcible entry tool
5. Thigh-high protective rubber boots
6. Pike pole
7. Air mask (connects to tank worn on back; positive pressure mask rated at 30 minutes)
8. Handie-Talkie speaker/microphone

ALL GEAR IS OSHA APPROVED

A heavy blow is dealt to a major fire by delivering thousands of gallons of water from the ladder pipe. The nozzle is affixed to the tip of the ladder, and its direction is controlled from the ground by means of halyards.

*A truckee climbs the aerial ladder to ventilate the upper-floor windows.*

Truckmen scurry to the roof to cut holes and ventilate the building. The faster the ventilation is completed, the sooner the building cools down, making it bearable for the men of the engine company to move in.

A ladder company uses ladder p[...]
cool a heavily involved cockloft fir[...]
and allow other truckees to enter [...]
search for victims.

A fire fighter attempts to gain entry to
search for and rescue trapped victims
in spite of the deadly heat and smoke.

A ladder pipe in operation at a
Harlem tenement fire.

*In the fire-filled night sky, the moon vies for attention.*

Fire illuminates the smoke-filled sky, providing a
dramatic backdrop, but for a truckee with a ladder
pipe it's business as usual.

A desperate measure—a truckman bridges two buildings with a ground ladder to attempt a rescue.

An acetylene torch is used to cut bars on windows to allow fire fighters entry.

A fiercely burning building awaits arrival of fire fighters.

*A long vigil on a cold night*

A fully involved factory—probably torched by an arsonist—
presents an arduous battle for arriving fire fighters.

**Above:** *An underground electrical fire fills the streets of New York City with fire-fighting apparatus.*

**Top left:** *Proper positioning of fire apparatus at a fire scene is essential.*

**Bottom left:** *The street is crowded with fire fighters awaiting their assignments at a multiple-alarm fire.*

*Flames shoot from the windows of this vacant tenement.*

*All hands are engaged at this downtown store fire. Interior as well as aerial methods of attack are used to extinguish the blaze.*

*Smoke-filled streets reduce visibility and bathe the apparatus in an eerie glow.*

*Thick smoke billows from this heavily involved two-story building.*

*Late into the night and long after the fire is extinguished, fire fighters wet down the scene of this multiple-alarm fire.*

The versatile tower ladder, or aerialscope, has the ability to deliver heavy-caliber streams of water on the fire while constantly moving or while remaining stationary at the area help is most needed. The ability to sweep a fully involved building allows it to cool the structure rapidly.

**Above:** *The ultimate aerial attack—a tower ladder and ladder pipe deliver a lethal blow to the Red Devil.*

**Top left:** *The sky bucket allows fire fighters a closer look while surveying a structure for hidden pockets of fire.*

**Bottom left:** *Fire fighters deliver the initial blow from the bucket to a raging, wind-fed fire.*

Stalemated by a three-alarm fire, the tower
ladder attempts to hold its ground.

No structure is safe from the Red Devil.
This church was gutted by the ravages of fire.

The dexterity of the tower ladder enables it to operate from under elevated subway tracks in New York City.

A fully involved four-story frame building poses
extreme danger to engine men stretching lines

A fire fighter blasts a top-floor fire with a heavy stream from his deck gun.

Air masks in place, fire fighters stretch a line into an upper-floor bedroom.

Two engine companies battle to gain entry into a burning two-story building.

*Having exhausted his thirty-minute supply of air, this fire fighter takes delivery of a fresh tank.*

*Rather than snake their hose up stairs and through hallways, this company takes the line up and over the fire escape and enters the burning building through a window.*

*The first-arriving engine company is confronted with an inferno. There is little doubt in their minds that heavy-caliber streams of water will be necessary to douse this fire.*

Battling this blaze requires the big guns. The heavy flames make interior fire fighting too dangerous.

Top left: *Gaining access to a fire can be difficult. Here a brick wall had to be breached to find the Red Devil.*

Bottom left: *Special hazards exist in a warehouse fire. Unknown materials, often stacked to the roof, pose unforeseen dangers.*

Right: *When the battle is over, fire fighters must search out and extinguish hidden pockets of fire.*

Below: *This odd-looking fire engine delivers large quantities of foam into inaccessible areas such as this store basement. The foam smothers the fire by depriving it of oxygen, which is far safer than sending men into the burning cellar.*

After approximately ten weeks of
intensive training, graduating
fire-company recruits demon-
strate their new skills at gradua-
tion ceremonies.

Right: The lighter side of training—a mock victim is rescued by
a would-be fire fighter at gradua-
tion ceremonies.

*Overcoming their fear of heights, trainees work out on an aerial ladder.*

*Recruits learn proper handling of tools from an experienced instructor.*

*Safe driving skills are taught to all recruits.*

New fire fighters are exposed to extreme heat and smoke. They must extinguish this fire in an open oil pit as part of their training.

Different fuels require different fire-fighting techniques. Here recruits battle a propane-gas fire. The crouching fire fighter must turn off the gas at its source while his comrades cool the flames with hose lines.

*Gaining entry in search of fire victims is the first priority upon arrival.*

*Fire fighters derive great satisfaction from a lifesaving rescue.*

*Doing it the old-fashioned way, a fireman uses a conventional yet effective ax to ventilate this burning house.*

*Modern power saws are used to ventilate this apartment house.*

Often, fire-charged rooms are ventilated by breaking windows with hand-held pike poles and hooks.

Butcher, baker, even your next-door neighbor—at the sounding of the alarm all are transformed into fearless fire fighters. This raging warehouse fire is tamed by an all-volunteer force.

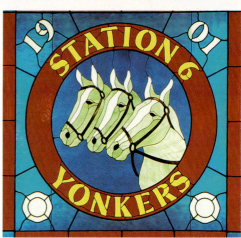

Top: Between alarms, fire fighters sharpen their artistic skills. Artist/fire fighter Bill Bresnan puts the final touches on another masterpiece.

Above: Stained-glass artist/fire fighter John Cuniffe's contribution to his station house.

Left: Chief John B. Stewart of Hartford, Connecticut, with fire fighters Zandra Clay and Maria Ortiz.

Fire fighters' gear is given a
chance to dry between alarms.

## Treasures from the collection of famed memorabilia buff James Piatti

Left: *A glass ceremonial trumpet*

Below left: *A chief's lantern*

Below: *Leather fire buckets were used until the early 1800s as the only means of supplying water to engines.*

Top and center right: *Vintage helmets and rare leather front pieces*

Bottom right: *Antique parade hats, or "stovepipes," made of pressed felt were used as a means of identification until the mid-1800s.*

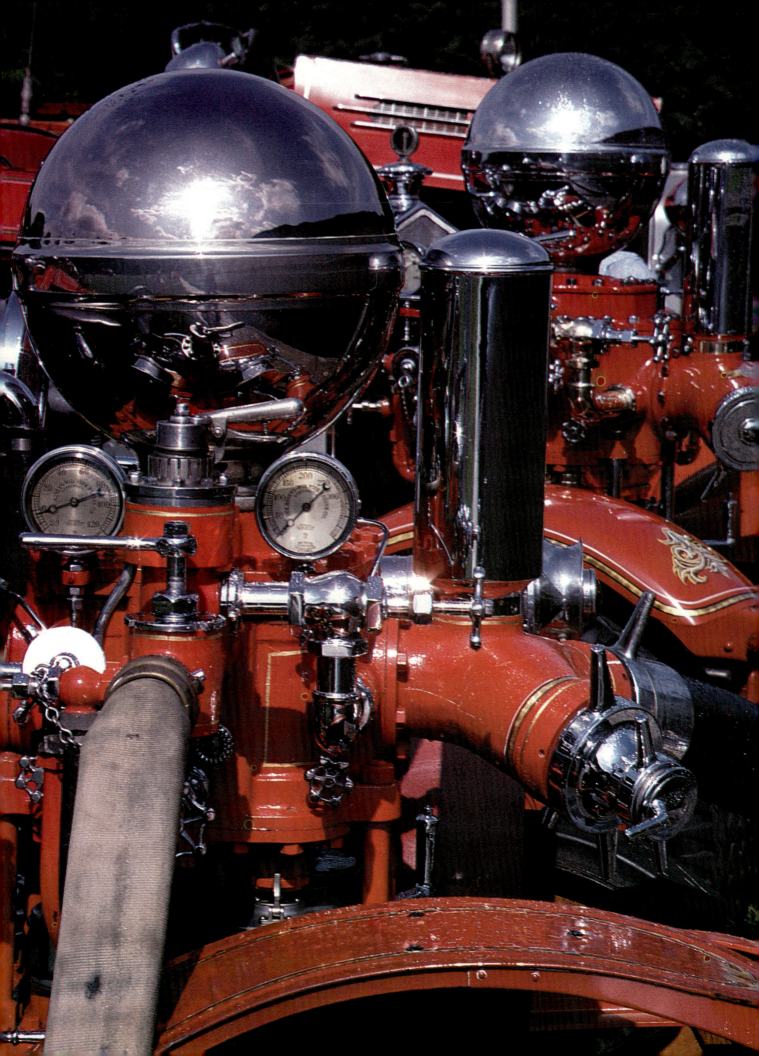

**Left and below:** *The distinctive profile of the Ahrens-Fox showing its stuff*

**Right:** *The Ben Franklin Bridge is the backdrop for an antique steamer at a Philadelphia muster.*

*Classic Mack bulldog*

*Polished to a high sheen, antique rigs are stunning to look at.*

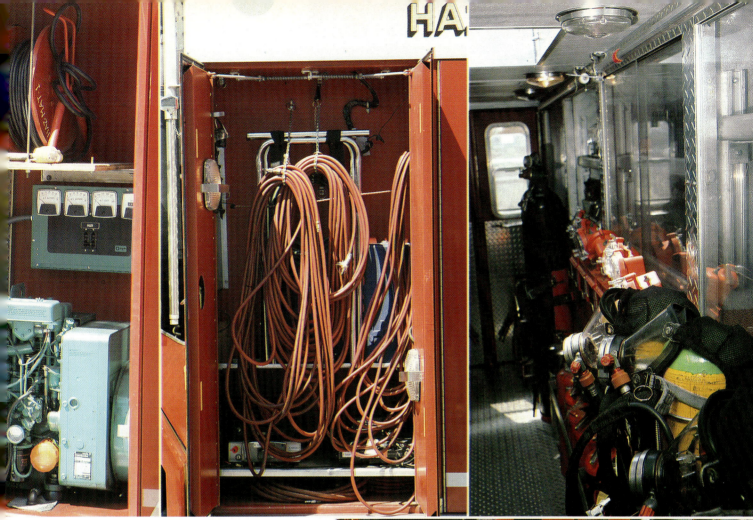

The hazardous-materials truck is equipped with hundreds of tools and gadgets for detection, identification, and removal of even the most toxic substances. When burned, simple everyday items such as plastics give off deadly cyanide gas. Liquid chemicals, when left to crystalize, become lethal explosives. On board the "haz-mat" rig, fire fighters consult a computer for the latest information on the handling and removal of the potential killer. To ensure uninterrupted communication, a modular telephone is on hand to enable the haz-mat team to establish a direct link to chemical experts and manufacturers nationwide.

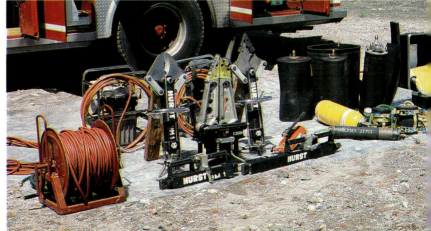

**Left:** *One of several protective suits available to keep fire fighters safe while performing their duties. These suits are airtight and made of special inert materials that are unaffected by corrosive chemicals.*

**Below:** *The haz-mat truck carries an ample supply of self-contained breathing apparatus, which allow fire fighters to work for extended periods of time in the midst of a chemical cloud.*

**Top right:** *Open compartment doors expose an array of haz-mat equipment.*

**Bottom right:** *One of New York City's experienced haz-mat teams*

*A status map of Brooklyn, New York, indicates fire companies available or operating at fire incidents. Color-coded lights show the status of each company.*

Always heard but rarely seen, the dispatchers assign the companies to all alarms of fire. Dispatchers are the vital link between imperiled victims and rescuers.

Both an engine and truck turn out at high speed as an alarm is sounded.

*The buck stops here. In command at a major blaze, a deputy chief communicates with his troops via walkie-talkie.*

The assistant, or battalion chief, is the first-line commander upon arrival at a working job. At a big burner, even the BC is not exempt from "pushin' in."

At a multiple-alarm fire, several chiefs often respond. This chief has been assigned a specialized task: communications coordinator.

Thoughts of tactics and strategy race through the mind of this battalion chief as he arrives at a smoky burner.

Tired and dirty, this fire fighter looks forward to being relieved.

Even in the heat of the battle, this fire fighter manages a friendly smile.

On a hot August afternoon or a frigid January morning, the fire fighter is hard-pressed to find comfort and protection from the elements.

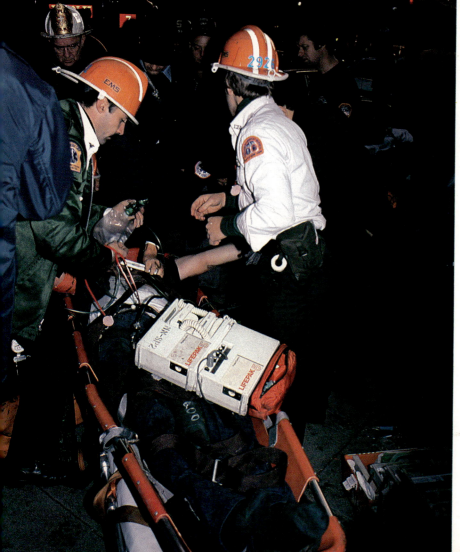

The only break for this lieutenant is a brief moment to change his air packs.

Danger abounds and often the enemy prevails. Severe lacerations, fractures, and smoke inhalation are some of the perils that confront a fire fighter every day at every fire.

*Sadly, at times the Red Devil deals a fatal blow and a comrade is lost. The death of a fire fighter is mourned by fellow fire fighters everywhere.*

F.D.N.Y.

TO THE TWELVE HEROIC FIREFIGHTERS WHO
PERISHED IN A FIRE IN MANHATTAN AT 23RD STREET
AND BROADWAY ON OCTOBER 17, 1966, AND ALL OF
THEIR BRETHREN WHO HAVE PERISHED IN THE LINE
OF DUTY, WE ADD THEIR LIVES TO THE ROLLS OF
VALIANT FIREFIGHTERS WHO HAVE MADE THE
SUPREME SACRIFICE

"GREATER LOVE THAN THIS NO MAN HATH
THAT A MAN LAY DOWN HIS LIFE FOR HIS FRIEND"
JOHN, CHAP. XV., V. XIII

DEPUTY CHIEF THOMAS A. REILLY
BATTALION CHIEF WALTER J. HIGGINS
LIEUTENANT JOHN J. FINLEY
FIREFIGHTER JOSEPH PRIORE
FIREFIGHTER JOHN G. BERRY
FIREFIGHTER JAMES V. GALANAUGH
FIREFIGHTER RUDOLPH F. KAMINSKY
FIREFIGHTER JOSEPH KELLY
FIREFIGHTER CARL LEE
FIREFIGHTER WILLIAM F. McCARRON
FIREFIGHTER BERNARD A. TEPPER
FIREFIGHTER DANIEL L. REY

MAY THEIR SOULS REST IN EVERLASTING PEACE

UNIFORMED FIREFIGHTERS ASSOCIATION
LOCAL 94—I.A.F.    NEW YORK CITY

*This monument stands in commemoration of the loss of twelve men at a single blaze in New York City.*

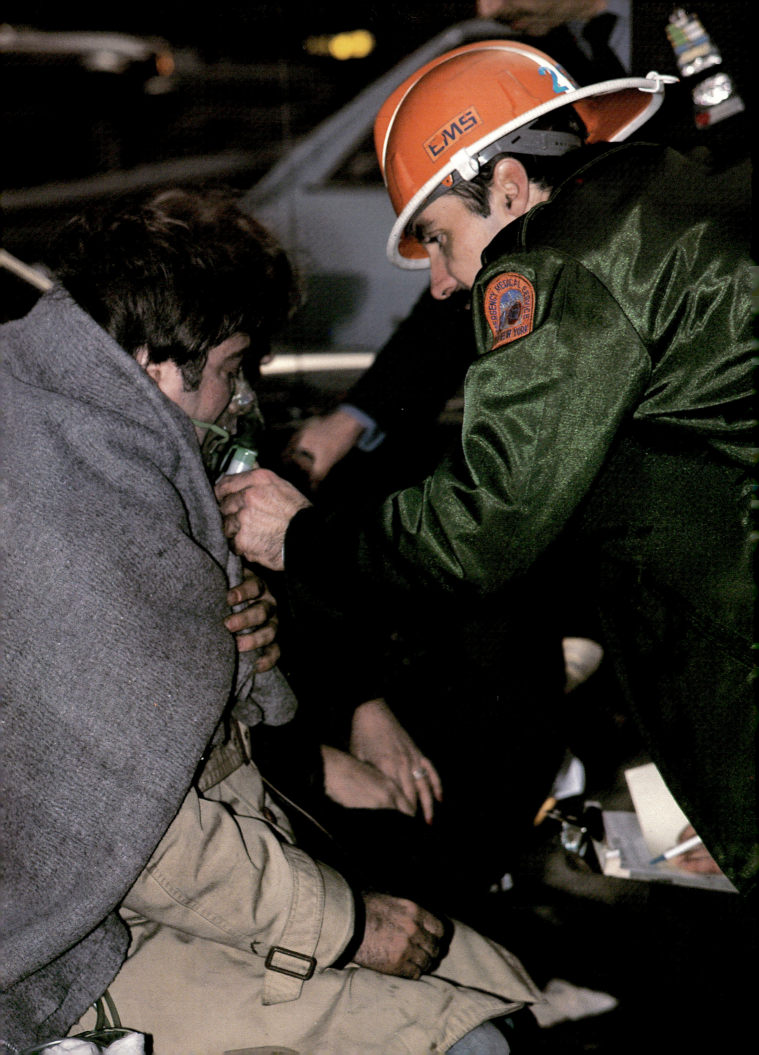

*A raging roof fire seems to dwarf these fire fighters.*

*Fire victims often lose their homes and belongings, escaping only with their lives.*

This massive deck gun mounted aboard an FDNY satellite is capable of directing 4,700 gallons of water per minute into an inferno.

The first-arriving ladder company prepares a ladder pipe for an aerial attack.

Right: Fire fighters prepare to ascend in the bucket from the foam-filled street.

*A shiny, horse-drawn steamer shows it still has "the right stuff."*

**Left:** *Jim Atkinson's dalmatian stands at the ready on an antique rig.*

**Below:** *Antique rigs on display at the American Museum of Fire Fighting in Hudson, New York.*

**Right:** *From 1965 to 1982 the super pumper was the world's largest fire engine. Now retired from service, the pumper has been replaced by the maxi-water system.*

SUPER PUMPER SYSTEM
SUPER PUMPER I
· F.D.N.Y.

*The main event, the evening meal, is frequently the victim of an untimely alarm.*
**Photograph by George Hall**

**Top right:** *Company emblems exhibit the personalities of individual units.*

**Bottom right:** *New York's field-communication unit handles hundreds of radio messages at the scene of all major fires. Photograph by Joe Pinto*

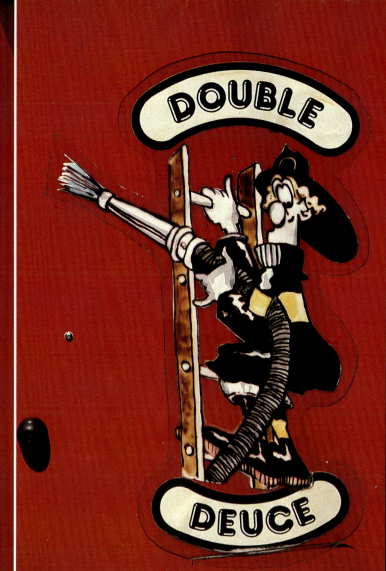

*Antique beauties strut their stuff.*

Fire fighters are off the mark in friendly competition at a fun-filled fire muster.

*Shooting the barrel in hose competition—all in fun*

*Every detail is considered
in the restoration of vintage rigs.*

A modern aerial ladder

A modern pumper

A modern aerialscope, or "tower ladder"

The Christmas spirit prevails as
an engine company "turns out."

An adjustable deluge set

Many large waterfront cities use fireboats. When major pier fires occur, only the fireboat can maneuver into position and then provide the unlimited water required to put out these fires.

An electrical transformer is cooled by fire fighters. Foam is used to extinguish fires in these large plants.

Manifolds can supply six 2½-inch lines at the same time.

Fire fighters stretch line into a building to prevent a fire from spreading.

*Facing extreme danger, the fire fighters close in on the enemy.*

Left: *Now that the fire is out, the cleanup begins.*

*Some like it hot.*

The cab of the rig is a welcome respite from the winter cold.

*Large fires draw huge crowds in the Bronx.*

*A fire truck awaits the return of a weary crew.*

A seemingly minor fire, such as a burning vehicle or a brush fire, presents dangers of its own—fuel tanks and tires could explode at any moment, and high winds can spread brush fires with alarming speed.

*After the fire is extinguished, a search must be made for any hidden smoldering embers. Overhauling, as it is called, is a tedious and tiresome task.*

*After hours of fighting a fire, where temperatures exceed 600°F, fire fighters are faced with ice-coated gear and frozen apparatus.*

# GLOSSARY

**Aerial** A 100-foot hydraulically powered ladder. Other trucks are tower ladders with a bucket mounted at the end of a telescoping boom.

**All hands** A fire incident at which the entire first alarm assignment is working.

**Backdraft** A sudden intensifying of the fire when an inrush of oxygen is introduced, perhaps by a window or door being opened. An often deadly threat. Also called an oxygen or smoke explosion.

**Battalion** A group of engine and ladder companies covering a given geographical district.

**Bottles** Refillable air containers for the Scott Air-Pak breathing apparatus.

**Charged line** A hose line with water coursing through it.

**Cockloft** Common attic space under the roof. Fire reaching the cockloft often spreads laterally with devastating results.

**Engine** Fire-fighting apparatus that pumps water and carries hose.

**Exposure** Surface on adjoining building that may be able to catch fire. A building adjoining a fire building is referred to as an exposure building.

**Halligan** The all-purpose steel prying tool invented by FDNY officer Huey Halligan.

**Handie-talkie** The trade name of Motorola's portable, two-way VHF radios, carried by key personnel at fire scenes.

**Jaws of life** More properly, the Hurst Forcible Entry Tool. This machine can cut or pry apart almost any material, and is most commonly used to extract injured motorists from wrecked vehicles. The tool is carried by all rescue companies. It features a hydraulic pump powered by a small gasoline engine.

**Ladder pipe** High-volume nozzle mounted atop an aerial ladder and controlled from the ground by lanyards.

**Mask** Refers to the mask plus the regulator and tank of the Scott Air-Pak. It gives the fire fighter up to 30 minutes of cool compressed air in a backpack tank.

**Master stream** Water at high volume and pressure, directed through outside nozzles. Usually employed when inside fire-fighting efforts have proved unsuccessful or unnecessary.

**Nob** The business end of an attack hose-line. Used regardless of the specific type of appliance employed: fog nozzle, brass tip, and so on.

**Overhauling** The difficult and dirty job of tearing apart walls, ceilings, and floors in the fire building to look for lurking hot spots.

**Tiller** The steer-from-the-rear position on tractor-trailer ladder trucks, manned by specially trained tillermen.

**Turnout** Old-time fire fighters responded to the call "Turn out!" Today the term refers to the rubber or Nomex overcoat worn in proximity to the fire.

**Working fire** A true fire incident, with obvious or visible flame.